# DC
# Locomotives
# of
# Indian Railways

Er. Twahir Alam

Copyright © 2018 Er. Twahir Alam

All rights reserved.

ISBN: 9781729018224

# DEDICATION

I humbly dedicate this book to all the hard working men and women of the Indian Railways, for it is their hard work and motivation that has kept the Indian Railways running.

# CONTENTS

|    | Acknowledgments | i  |
|----|-----------------|----|
| 1  | Introduction    | 1  |
| 2  | DC Traction     | 3  |
| 3  | DC Locomotives  | 5  |
| 4  | WCP Series      | 6  |
| 5  | WCG Series      | 10 |
| 6  | WCM Series      | 15 |
| 7  | WCU Series      | 22 |
| 8  | YCG Series      | 23 |
| 9  | Virar Dead Zone | 25 |
| 10 | Conclusion      | 26 |

# ACKNOWLEDGMENT

I would hereby like to thank my wife, Sabina Sultana for her support during writing of the book.

# INTRODUCTION

The Indian Railways is righty regarded as the lifeline of the nation as it is the true mover of India. Its journey started on a humble note on the 16th of April 1853, with a short train ride from Bombay's Bori Bunder railway station to Thane, a distance of about 34 kms. The train carried 400 passengers in 14 carriages which were pulled by 3 steam locomotives. The Railway has grown leaps and bounds and has grown to become the fourth largest railway network by size in the World, with total track of 67,368 kms. It now operates about 13,000 passengers trains in its network of about 7,349 stations and also, about 9,200 goods trains daily in its network.

The evolution of the Indian Railways is one of the most interesting episode of modern Indian history but sadly one which has received very little attention. Yet for technologists, engineers and railway enthusiasts like me, it is a breath taking journey and one which has to be studied in detail and told to the World.

The journey and growth of the Indian Railways has not been a smooth or linear one, but one filled with success, failures and lots of experimentations. It is this which makes the study of its history even more fascinating. Thus, we can find many technologies and policies being used and tried to make the Railways more efficient and their success has contributed greatly to the growth of Indian Railways. Yet many a times, we have seen that many such experiments have failed or that new developments have rendered them to be obsolete. In this book, we would look into one such technology which was greatly used by the Indian Railways and had given it lot of success but due to advancement of technology, it got lost in the sands of time.

The technology which we are talking about is DC traction technology and it was the first electric traction technology used in India. DC power was not only first used for traction purposes but also was the first distribution system used all over India but due to rapid development of AC technology, the use of DC power was gradually set aside.

The research conducted by the French Railway (SNCF) concluded that 25 KV AC as the most economical for Railway purposes. So, the Indian Railways in 1957 decided to adopt the same technology as its standard with SNCF as its consultant. This relegated the DC traction use to only the Mumbai section. It is here where it did survive for few decades but with the decision to convert the whole nation to a 25 kV AC network in 1996-97, mainly due to limitations and more maintenance costs of DC systems.

It was in the budget of 1958-59 which was announced by Shri Jagjivan Ram on the 17th of February, 1958 that it was decided that the 25 KV AC system would be used as the standard for electrification works in India. It also announced that SNCF would be appointed as the Technical Associates to advice on all technical problems connected with electrification under 25 KV AC system, and also to supervise the actual execution of the Projects.

The conversion from DC to AC was completed by Western railway by 2012 and by Central Railway by 2016. Thus, the history of use of DC traction in India came to an end, even though it is still used in the Metro systems and in the tram system in Kolkata.

Today, DC traction is no longer used by Indian railways but today we would glance back into the pages of history of its heyday when they ruled a part of the Indian Railways networks. Here, we look at the workhorses of the DC networks which used to pull the trains.

# DC TRACTION

Today about 33% of the Indian Railways tracks are electrified and electrification is regarded as a step forward. So, new demands are seen everywhere to covert the lines which use diesel traction to electric traction. Due to the economic feasibility and environment benefits, the Railways has announced on 22nd March 2017, that the entire railway network in India would be electrified by 2022.

Given the euphoria about electric traction, it may come as a surprise that electric traction predates the diesel traction and was put into use much earlier than the diesel technology but the only difference was that the technology used was DC and not the AC traction we see today.

The first DC propelled train ran between Bombay Victoria Terminus and Kurla on the Harbour Line, on 3 February 1925 on the Great Indian Peninsula Railway (GIPR). Electrical technology was introduced to tackle the heavy gradients of the Western Ghats. The technology was based on 15 kV DC. The success of their introduction, led to DC lines being introduced in other places in India also. On 5th January 1928, the suburban section between Colaba and Borivali of the Bombay, Baroda and Central India Railway, was electrified with the same technology. On 11 May 1931, the same technology was used to electrify the section between between Madras Beach and Tambaram of the Madras and Southern Mahratta Railway.

At the same time, a new technology based on 30 kV DC, was used during the electrification of the Howrah-Burdwan section of the Eastern Railway in 1958. The first EMU service began on the Howrah-Sheoraphuli section on the 14th of December 1957. With it, DC traction reached its peak as it was in 1957 only that the Indian Railways decided to go with the findings of SNCF and shifted to the use of AC traction as its standard.

It was in the railway budget of 1997-98, which announced by the then Railway Minster, Shri Ram Vilas Paswan, on the 26th of February, 1997, that as the existing 1500 volt DC system in Mumbai area had reached its operational limits and was causing a serious constraint in handling any additional traffic. The existing 1500 volt DC traction system on both Central and Western Railways would be converted toconverted into 25 kv AC single phase 50 Hz system.

Thus, the decline of DC traction began in India but not before they had rendered great service to the railways and to the nation as a whole. The engines were greatly respected by all and they were not only a product of engineering but also a work of art. While they are not able to generate the as much nostalgia and romanticism as the steam locomotives have done. So as such their retirement has not been highlighted as much as the steam locomotives were but they too remain an important part of the history of Indian Railways and one which we must not allow to get lost in the pages of history.

# DC LOCOMOTIVES

The DC locomotives used by the Indian Railways in its main lines are as follows:-

Passenger locomotives:-
1. WCP 1 & WCP 2.
2. WCP 3 & WCP 4.

Goods Locomotives:-
1. WCG 1.
2. WCG 2

Mixed type locomotives:-
1. WCM 1.
2. WCM 2.
3. WCM 3.
4. WCM 4.
5. WCM 5.
6. WCM 6.

Electric multiple units (EMU):
1. WCU 1 to WCU 15.

Meter gauge
1. YCG-1.

# WCP 1

These were the first of the electric locomotives to be used in India and were imported in 1930s. They were designed as per the GIPR's tenders for high-speed Broad Gauge 1500 V DC locomotives for service on the Mumbai-Pune and Igatpuri routes. It is said that GIPR wanted speeds of 85 mph (137 kph) but that this was not possible at that time. The EA/1 was thus selected for mass production and WCP1s hence turned out to be the first electric locomotives in India.

They were produced by Vulcan Foundry, UK and Swiss Locomotive and Machine Works (SLM) with electrical components being supplied by Metropolitan Vickers, UK. The first one was built in 1928 and 21 more were built in 1930. They were classified as EF/1 but later reclassified as WCP 1.

They had a wheel arrangement similar to the steam locomotive type. According to the NRM booklet, the EA/1s had a rigid wheelbase of two driving wheels. The third driving wheel is articulated with the third carrying wheel. Each of them was powered by two traction motors generating 350 hp each connected to the axle through a Universal Drive. This in turn could be connected in various combinations to give six different speeds.

Also, it was these electric locomotives which heralded the story of high speed trains in India, as they used to complete 192 km steeply graded Bombay-Poona journey with the 7-car Deccan Queen in just 2 hr. 45 min. in the 1930s but today the fastest train on the route does the run in about 3 hr. They remained in service till the 1980s.

Sir Roger Lumley:-

The first one of the class was rechristened "Sir Roger Lumley", after Sir Roger Lumley who was Governor of Bombay from 1937 to 1943. Under the GIPR it had the serial number EA/1 4006, which later became the WCP/1 20005 under the Central Railway. The same is now under display at the National Rail Museum at Delhi.

It is believed that the name 'SIR ROGER LUMLEY' which this engine now bears was actually applied to another locomotive # 20024, of a subsequent and more powerful class WCP/2. This sometimes leads to confusion in the minds of purists regarding the actual class of this locomotive. Name notwithstanding, the engine at the National Rail Museum is engine is a WCP/1, and not a WCP/2 as some might be led to believe.

One more EA/1 is preserved in the Nehru Science Centre in Bombay.

Data Sheet:-
- Manufacturer:- Vulcan Foundry, UK & Swiss Locomotive and Machine Works (SLM)
  - Production Period:- 1928 – 1930.
  - Numbers Produced:- 22.
  - Wheel Arrangement:- 2-Co-1.
  - Rated Power Output:- 2160 hp.
  - Top (Rated) Speed:- 120 kmph.
  - Weight:- 102-105 Tonnes.
  - Starting TE:- 15240 kg force.
  - Serial Numbers:- 20002-023.

# WCP -2

The lone WCP2 was produced in 1938, eight years after production of the WCP1 had ceased. It was then classified EA/2 by GIPR and later as the WCP 2 by Indian Railways. It is not clear why it was given a distinct classification as it is in essence the very same as the EA/1 and shared all its specifications. It was numbered #4025 by the GIPR and later #20024.

There is a belief that it was this #20024 that hauled the Deccan Queen on its first run in 1930, which is wrong. It was the then EA/1 #4006 (Sir Roger Lumley), later renumbered #20004 that did those honors. In fact the engine was not even produced when the Deccan Queen started running.

Fact Sheet:-
- Manufacturer:-
- Production Period:- 1938.
- Numbers Produced:- 1.
- Wheel Arrangement:- 2-Co-1.
- Rated Power Output:- 2160 hp.
- Top (Rated) Speed:- 120 kph.
- Traction Motors:- 6.
- Weight:- 102-105 Tonnes.
- Starting TE:- 15,240 kg force.
- Serial Numbers:- 20024.

# WCP 3 & WCP 4

These were also one of the earliest used electric locomotives in India and were used by the Great Indian Peninsular Railway (GIPR). They were built by Hawthorn Leslie and Company, which is based in Hebburn, UK and only one of each class was built and imported to India in 1928. They had a 2'Co2' wheel arrangement.

EB/1 and EC/1 were two more prototype locomotives that GIPR had procured along with the EA/1 as a competing bid to the tender put out by GIPR but lost out to SML & Metro Vickers. They were manufactured by Leslie Hawthrone and Co. in UK, but with two different types of drives coming from General Electric and Brown Boveri respectively, attempts at hitting the 137 kph speeds as requested by GIPR. They were more powerful and had a symmetrical wheel arrangement of 2-Co-2 which again, looks like was inspired by steam locomotives.

EB/1 was #4001 and EC/1 was #4002 but these were not selected for further mass production. They remained in service with the Indian Railways until the 1960 when they were assumed scrapped. Though largely forgotten, the WCP3 and WCP4 were still only the second and third electric locos in India, and largely share the WCP1's legacy.

Fact Sheet:-
- Manufacturer:- Hawthorn Leslie and Company.
- Production Period:- 1928.
- Numbers Produced:- 1 each.
- Wheel Arrangement:- 2-Co-2.
- Rated Power Output:- 2250 hp (WCP 3) 2390 hp (WCP 4).
- Top (Rated) Speed:- 120 kph.
- Traction Motors:- 6 each.
- Weight:- 113 Tonnes.
- Starting TE:- 10,890 kg force.
- Serial Numbers:- 20001 (WCP 3) & 20002 (WCP 4).

# WCG 1

The WCG 1s were the first of electric locomotives to be imported to India, which was solely dedicated for freight purposes. They were supplied to the GIPR in 1928 and were intended for use on the Bombay-Pune section for pulling the heavy freight trains. This was due to the steep graded ghat sections which are present between Kalyan and Igaturi and between Kalyan and Pune. Coupled with the increased traffic in this line, it was difficult to service the line with steam locomotives.

A total of about 41 were inducted into service, while the first ten were made by the Swiss Locomotive Works, Winterthur, and remaining thirty one by the Vulcan Foundry (with electricals from Metropolitan Vickers Electrical Company Limited). They were given serial numbers from 20025 to 200065 and classed as EF/1 under the GIPR which was later changed to WCG-1, under the Central Railway.

They had four 650 hp motors (total power often quoted as 2610hp), driving two three-axle bogies through connecting rods. They had an articulated frame, suitable for rounding the sharp bends on the ardous hill route. The engines are styled around the renowned Swiss 'crocodile' class of engines, so called due to their low slung profile and very long wheelbase, and an alleged resemblance to that animal while rounding bends. The pantograph was manipulated by a pole inside the driver's cab, just behind the driver. Several of them had steam-locomotive type of whistles which were later replaced by electric horns.

Their unusual features included an articulated body (made them ideal for use in heavily curved ghat sections). They also featured regenerative braking. They were known for their superior tractive characteristics on the ghat sections; however, the exposed link mechanisms had to be oiled very frequently in all kinds of weather.

Locally they were known as "khekda" ("crab"), as they made a curious moaning sound when at rest, and while on the run an unusual swishing sound from the link motion can be heard.

The engines were greatly admired by their drivers and crew and retired from active service i.e. from main line workings in 1974 .They were later used as bankers on the Karjat-Lonavla section, and were later used as shunting locomotives till as late the 1992 at Bombay VT and Lonavla. The use of these locomotives even after almost six decades of service is an excellent example of their capacities and also of Swiss Technology.

Today only a few of them remains, while one is in exhibit at the National Rail Museum at Delhi, the others are at the Wadi Bunder loco trip shed.

Fact Sheet:-
- Manufacturer:- Swiss Locomotive Works, Winterthur, and Vulcan Foundry.
- Production Period:- 1928-29.
- Numbers Produced:- 41.
- Wheel Arrangement:- C+C.
- Rated Power Output:- 2600-2890 HP.
- Top (Rated) Speed:- 72-80 kph.
- Weight:- 125 Tonnes.
- Starting TE:- 30,4800 kg force.
- Serial Numbers:- 20025 to 20065.

## Sir Leslie Wilson

The first of the class of locomotive which was imported to India was rechristened as "Sir Leslie wilson", who was the Governor of Bombay from 1923 to 1926. It was he who had flagged off the first electric train in India on the 3rd of February 1925. It ran from Bombay VT to Coorla on the harbor line, a distance of about 16 kms with a impressing speed of 50 miles per hour.

This locomotive bearing the number EF/1 4502, later WCG/1 20027 is on display at the Rail Museum in Delhi and thus, their contribution to Indian Railways has been preserved for everyone to see. It has also been reported that the name 'SIR LESLIE WILSON' originally borne by this locomotive was originally the name of another locomotive # 20025 (EF/1 4500), which was also a WCG/1, and which was in fact the first engine of this class.

# WCG 2

These locomotives were custom-built for freight for the 1.5 KV DC section of the CR Mumbai Division by Chittaranjan Locomotive Works in 1970. A total of fifty seven was produced by them and production continued till 1977. Most of them were used under the Mumbai division. Infact, they were used extensively for passenger services also, during the 2000s when huge shortage was faced in this section.

They had a rated power of about 4,200 hp (3,100 kW), with maximum speeds of upto 90 km/h (56 mph) and a tractive effort of 35600 kgf of tractive effort. They employed a three series-parallel motor combinations with weak field operation. The bogie design was finalized and designed by the manufacturers. They were ideally suited for multiple operations and up to 3 units were used. Some units of the WCG-2 model have a different gearing ratio for banking duties and are classified WCG-2A.

This loco has a very loud noise caused by the blowers used to cool the dynamic brake resistors. They were normally coupled in pairs or triples, to haul freight trains in the Bombay - Igatpuri/Pune sections. They are also used as bankers on the ghat sections.

There was a period around 1992 -1996 when the Mumbai division was desperately short of motive power due to the aging and failure prone WCM fleet. The punctuality of trains in and out of Mumbai went greatly affected due these failures. During this period the WCG-2 was used on many passenger trains but the 'Deccan Queen' has been hauled only once by a WCG-2 and only a few times by a WDM-2 when its power, the WCM-1, failed.

The Ghat banking duties in the Bhore ghat (Karjat - Lonavala) and Thull ghat (Kasara - Igatpuri) were exclusively handled today by them and on some occasions some express trains are hauled. When speed restrictions of 90km/h were imposed on the Mumbai (CSTM) - Igatpuri route, even superfast trains can be hauled on this section by them. They shared bogies with the WAM-4, WCAM-1, WCAM-2, WDM-2/2C, WDS-6, WDM-7 locos (Alco type cast trimount (Co-Co) bogies). Their use was later restricted to performing banking duties only. The locomotives have now been fully withdrawn service.

Fact Sheet:-
- Manufacturer:- Chittaranjan Locomotive Works.
- Production Period:- 19270-76.
- Numbers Produced:- 57.
- Wheel Arrangement:- Co-Co.
- Rated Power Output:- 4,200 HP.
- Top (Rated) Speed:- 80 kph.
- Weight:- 132 Tonnes.
- Starting TE:- 35,600 kg force.
- Serial Numbers:- 20104 to 20160.
- Traction Motors:- Heil TM4939AZ (690hp, 700V, 800A, 1070 rpm, weight 3670kg) Six motors, axle-hung, nose-suspended, force-ventilated. (4200hp total power, 1640 1-hour continuous rating in series mode.)
- Pantographs:- Two, Faiveley AM-18B.

# WCM 1

These locomotives were manufactured by English Electric / Vulcan Foundry and the auxiliaries were produced by Westinghouse. They were the first electrics with the now familiar Co-Co wheel arrangement to be used in India. They are characterized by their large size and unusually long hoods. The position of the entrance doors is also unusual, being not at the sides of the cabin, but through an entrance in the middle of the loco body side.

They were introduced in 1954 and several of them were rebuilt in 1968. They were used on superfast trains such as the Indrayani Exp. and the Deccan Queen until the 1990's. They were rarely used for freight. Air brakes were used for loco with regenerative braking. While vacuum brakes were used with the trains. Three different series-parallel motor combinations are available, as well as weak field operation. Multiple unit operation was not possible with them.

They have now been pulled out of active service and one is reported to have been sent to the National Rail Museum, and the other is now the main exhibit at the Chennai Rail Museum.

Fact Sheet:-
- Manufacturers:- Vulcan Foundry.
- Production Period:- 1954-55.
- Numbers Produced:- 7.
- Wheel Arrangement:- Co-Co.
- Rated Power Output:- 3,120 HP.
- Top (Rated) Speed:- 105-120 kph.
- Weight:- 124 Tonnes.
- Starting TE:- 31,300 kg force.
- Serial Numbers:- 20066 to 20072.
- Traction Motors:- 6 axle-hung, nose-suspended, force-ventilated English Electric 514/2C DC motors (615hp, 700V, 700A, 738 rpm, weight 3594kg).
- Gear Ratio:- 59:16.
- Pantographs:- English Electric, PNL4-F1. Two provided.

# WCM-2

The WCM-2s like the WCM-1s, were manufactured by English Electric / Vulcan Foundry with the auxiliaries being supplied by Westinghouse (compressor, etc.) and North-Boyce (exhauster). They were slightly smaller than the WCM-1, but had normally positioned entrance doors. They were initially built to run on the 3kV DC sections in the Calcutta area but were rendered obsolete when still quite new when the Calcutta area was converted to 25kV AC. The RDSO Lucknow modified them to work on 1.5kV DC without loss of power, and they were subsequently moved to the Bombay VT - Poona - Igatpuri area. A total of fifty seven of them were built in 1956-57 they remained in till the 1980s.

They were mostly used for passenger duties despite the mixed classification. They used the Three series-parallel combinations with weak field operation. Air brakes were used for locomotives and vacuum brakes for train.

Factsheet:-
- Manufacturers:- Vulcan Foundry.
- Production Period:- 1956-57.
- Numbers built:- 12.
- Wheel Arrangement:- Co-Co.
- Rated Power Output:- 3,120 HP.
- Top (Rated) Speed:- 105-120 kph.
- Weight:- 113 Tonnes.
- Starting TE:- 31,300 kg force.
- Serial Numbers:- 20175 to 20186.
- Traction Motors:- English Electric 531A (520hp, 1450V, 260A, 1165 rpm, weight 3445kg). Six motors, axle-hung, nose-suspended, force-ventilated.
- Gear Ratio: 62:16.
- Pantographs: English Electric PNL6-B1. Two provided.

# WCM-3

The WCM-3s were built by Hitachi with auxiliaries being supplied by Westinghouse and North Boyce. They were built in 1957-58 and were the smallest of the WCM series. They too were built for the 3kV Calcutta area and later converted to run on 1.5kV DC. Only three were these were ever built and bore the numbers from 20073 to 20075. All three of them have now been withdrawn from service.

The WCM-3 units were characterized, by separate light enclosures for the parking / marker lights (next to the headlight) and the tail lamps (just above the buffers). They were later used mostly for freight. They too used the Three series-parallel motor combinations, with weak field. The air brakes were for the locomotives while vacuum brakes for the train.

Fact Sheet:-
- Manufacturers:- Hitachi.
- Production Period:- 1958.
- Numbers built:- 3.
- Wheel Arrangement:- Co-Co.
- Rated Power Output:- 3,600 HP.
- Top (Rated) Speed:- 105-120 kph.
- Weight:- 113 Tonnes.
- Starting TE:- 28,200 kg force.
- Serial Numbers:- 20073 to 20075.
- Traction Motors: Hitachi HS 373-AR-16 (600hp, 1450V, 330A, 927 rpm) Six motors, axle-hung, nose-suspended, force-ventilated.
- Gear Ratio: 51:16.
- Pantographs: Two.

# WCM-4

The WCM-4s, were also built by Hitachi, with auxiliaries being sourced from Westinghouse and North Boyce. They were mostly built in 1960 and were larger and more powerful versions of the WCM-3 but with normal light enclosures. Initially they were used to haul superfast and other express trains, but later due to technical difficulties they were relegated to freight operations.

These are the only WCM series locos to be used almost exclusively for freight duties, despite the mixed classification). Several of them were fitted with CBC couplers and were the last imported engines to come with bonnets (noses) at either end. They too used the Three series-parallel motor combinations, with weak field operation. Air brakes and regenerative braking were used for the locomotive while vacuum brakes were used for train.

Only seven of these units were built.

Fact Sheet:-
- Manufacturer:- Hitachi.
- Production Period:- 1960.
- Numbers built:- 7.
- Wheel Arrangement:- Co-Co.
- Rated Power Output:- 4,000 HP.
- Top (Rated) Speed:- 105-120 kph.
- Weight:- 125 Tonnes.
- Starting TE:- 31,300 kg force.
- Serial Numbers:- 20076 to 20082.
- Traction Motors:- Hitachi HS 373-BR (675hp, 700V, 765A, 850 rpm, weight 4500kg). Six motors, axle-hung, nose-suspended, force-ventilated.
- Gear Ratio:- 73:16.
- Pantographs:-Two, type KP-120.

# WCM-5

The WCM-5s, were built first of the electric engines to be built in India and this was done by Chittaranjan Locomotive Works as per design and specifications of RDSO. The auxiliaries were supplied by Westinghouse and North Boyce. They were similar to the WCM-4 locomotives in traction motor arrangement, etc. The first was locomotive of this class was commissioned on the 14th of October 1961 and was named 'Lokamanya'. It was flagged off on that day by the First Prime Minister of India, Pt. Jawahar Lal Nehru.

In the WCM series, these are the first to use half-collector pantographs. There is a wide variation in the side window grille profiles, and very few of these units look alike. Several are nowadays fitted with CBC couplers. They too used the Three series-parallel combinations of motors, with weak field operation. The brakes used were Vacuum brakes and regenerative braking.

They were mostly used for passenger duties and now been withdrawn from service.

Factsheet:-
- Manufacturers: Chittaranjan Locomotive Works.
- Production Period:- 1961-63.
- Numbers built:- 21.
- Wheel Arrangement:- Co-Co.
- Rated Power Output:- 3,700 HP.
- Top (Rated) Speed:- 105-120 kph.
- Weight:- 124 Tonnes.
- Starting TE:- 31,300 kg force.
- Serial Numbers:- 20083 to 20103.
- Traction Motors: Hitachi HS 373-BR (675hp, 700V, 765A, 850 rpm, weight 4500kg).
- Gear Ratio: 59:16
- Pantographs: Two Faiveley AM28 BB

# WCM-6

The WCM-6 like the WCM-5s, were built in 1996 by Chittaranjan Locomotive Works as per RDSO's design and specifications but only two of them were ever built. They had underslung compressors and static converters were of Siemens make, while the compreesor was made by Elgi.

They were used for light freight duties, especially on the Kalyan-Karjat section. Only two of these were built (#20187, #20188), perhaps because CR preferred the WCAM-3 instead.

Ther were mostly used for light freight duties on the Kalyan-Karjat section. One of the WCM-6 was seriously damaged in a fire but it was restored by the Kalyan loco shed. They were also used for shunting duties around Bombay in the nineties but were later used to haul passenger trains in and around Bombay in early 2000s. They also may have been used for banking operations up to Lonavala.

They have high-adhesion bogies similar to those on the WAG-7 and were often coupled with WCG-2 locos. The Speed control was done by three series-parallel motor combinations with weak field operation. Air brakes were used for the locomotives while vacuum brakes for the train.

They have now been converted to AC only operation with WAG-7 specifications, but are still classified as WCM-6. They are now restricted to shunting/departmental duties only with a reduced of maximum speed of 65 kmph, as per provisional certificate issued by RDSO.

Factsheet:-
- Manufacturers:- Chittaranjan Locomotive Works.
- Production Period:- 1995.
- Numbers built:- 2.
- Wheel Arrangement:- Co-Co.
- Rated Power Output:- 5,000 HP.
- Top (Rated) Speed:- 105 kph.
- Weight:- 123 Tonnes.

- Starting TE:-
- Serial Numbers:- 20187 and 20188.
- Traction Motors:- Hitachi H5 15250. Axle-hung, nose-suspended, force-ventilated.
- Wheelsets:- High-Adhesion Co-Co fabricated bogies.
- Gear Ratio:- 18:64
- Pantographs:- Two, Faiveley AM-18B

## WCU 1 TO WCU 15

These were the earlier of the Electric Multiple Units (EMUs) to be used on the Indian Railways and were used mostly on the suburban routes of Mumbai. They were also used on the Lonavala-Pune section.

Most of the models had DC traction motors with rheostatic control (resistance banks to vary the input power supply).

# YCG 1

The YCG-1 was a 43 ton 1500 Volt DC locomotive and was imported to serve the Chennai area in the early 1930s. It was amongst the first to be used on this section and was the only DC meter Gauge locomotive to be used by Indian Railways. Four of these locomotives were imported from Hawthorn, Leslie & Co. with electrical equipment supplied by English Electric Co for Rs 89,928 each.

The two bogies are connected by a link and the 4 Axle Hung motors are connected in two groups of 2 motors each permanently in series and are started with all in series connection. The locomotive is provided with a transition system to change the motor circuit from series to parallel arrangement at high speeds. One outstanding feature of the YCG/1s was their capability to work on un-electrified yard lines on attachment of a trailer car housing 'ET' class 4-wheeled battery tenders. This 650 HP locomotive had a maximum speed of 65 km/h. These locos had a roughly rectangular, box-like body with a cab at either end. The two bogies had interconnecting linkages to allow easier negotiation of sharp curves. Two 'diamond' style pantographs were provided for current collection.

The YCG/1s were used to run on the suburban lines, and never had the privilege of working in main line trains. They were also used for switching operations and shunting freights between Madras Beach and Tambaram.

Two locomotives of this class is preserved as of now, one bearing the serial number 21900 is preserved at the National Rail Museum in New Delhi and can be found on a pedestal outside the Tambaram car shed in Chennai.

Fact Sheet:-
- Manufacturer:- Hawthorn, Leslie & Co.
- Production Period:- 1930.
- Numbers Imported:- 4.
- Wheel Arrangement:- Bo'Bo'.
- Rated Power Output:- 650 HP.

- Top (Rated) Speed:- 65 kph.
- Weight:- 43 Tonnes.
- Starting TE:-
- Serial Numbers:- 21900 to 20903.

## VIRAR DEAD ZONE

In 1957, when the Indian Railways decided to adopt the 25kV AC as its standard, the use of DC technology in India came to an abrupt end. In a few years, almost all the DC lines were converted to AC lines with the exception of Mumbai region and it remained as an island. Thus, it remained isolated from the entire Indian Railway network and continued to have separate technologies and stock. This led to this unique feature of this part of the railway network which is commonly known as the Virar Dead Zone.

The dead zone is an electrically neutral area at Km 63 on the Western Railway out of Mumbai. This zone allows the dual voltage locos of the Western Railway to change the working voltage while on the run. The area between Churchgate and Virar is charged to 1.5 KVDC while the area north of Virar is charged to 25 KVAC.

The zone is length of the overhead lines which is not supplied with electric power and lies between the AC and DC lines. This usually extends for a length of about two or three sections. About a kilometer before this dead zone, a sign alerts driver with a '1000 meters' warning followed by another for '500 meters' and then a sign saying 'Dead Zone'.

Going from Mumbai, the driver has to shuts down most of the equipment in the locomotive and then lowers the DC pantograph. The locomotive now has to travel past the dead zone without power. When the AC section of the line is reached, the raises the AC pantograph and soon the voltmeter shows 25 kV and he restarts the traction and other equipment.

The above maneuver is very tricky, as care has to be taken, to ensure that the correct type of power is selected and that the locomotive is enough momentum to sail through the dead zone. If the locomotive has less than the required momentum, it would stop dead in the neutral zone and would require another locomotive to dish it out. Also, a wrong selection of the type of power to be used would cause serious damage to the locomotive.

The dead zone was quite a unique feature and served as a

canal between the sea of AC network and the small pool of DC network. It led to the development of the unique dual voltage locomotives ie the WCAM series. They too served the Indian Railways very efficiently but the decision to convert the entire Indian Railways network to AC technology marked the end to it. In 2016, with the last part being converted, the use of DC technology in Indian Railways came to an end and also led to the death of the Virar Dead Zone. This relegated it to the pages of history. Today no locomotive is required to change its pantograph from AC to DC at the Virar Dead Zone but still whenever any locomotive crosses the dead zone, the maneuvers of the old days must flash in the minds of the people who have witnessed it.

# CONCLUSION

In the 2000s, the Indian Railways had decided that the entire electric rail network in India would be converted to AC and by 2016; the entire network was converted to AC. This ended the use of DC technology in the Indian Railways. By this time, most of the DC locomotives used had already been long retired from service and the few remaining ones were converted to AC. Even the dual technology locomotives were converted to run as fully AC locomotives.

Thus, ended the story and the contribution of DC locomotives in Indian Railways. While they are still used in the trams and metro of Indian Railways but they are no longer seen running on the Mumbai section. The heroes of the Mumbai-Pune and Igatpuri routes have been replaced but yet their memories still remain.

Er. Twahir Alam

SIR ROGER LUMLEY

## ABOUT THE AUTHOR

Er. Twahir Alam is a civil servant, working for the Government of Assam in Assam Civil Service. He is an Electrical Engineer having passed out from Assam Engineering College and has lots of experience working in Engineering field also, working as Executive for Crompton greaves in Delhi and for Assam Power Distribution Company Limited as a Assistant Manager. He also worked as a Excise Inspector in the Excise Department. He is a train enthusiast also and has a section of his youtube channel dedicated to Indian Railways.

www.ingramcontent.com/pod-product-compliance
Lightning Source LLC
Chambersburg PA
CBHW031558210526
45464CB00003B/1331